I0018282

Machine Learning

rick by rick

Epoch 1

Dmitry Vostokov

Machine Learning Brick by Brick, Epoch 1: Using LEGO® to Teach Concepts, Algorithms, and Data Structures

Published by OpenTask, Republic of Ireland

Copyright © 2020 by OpenTask

Copyright © 2020 by Dmitry Vostokov

All rights reserved. No part of this book may be reproduced, stored in a retrieval system, or transmitted, in any form or by any means, without the prior written permission of the publisher.

Product and company names mentioned in this book may be trademarks of their owners.

OpenTask books and magazines are available through booksellers and distributors worldwide. For further information or comments, send requests to press@opentask.com.

A CIP catalog record for this book is available from the British Library.

ISBN-13: 978-1912636501 (Paperback)

Revision 1.00 (April 2020)

Preface

My interest in artificial intelligence goes back to school years when I was intrigued by pictures of a perceptron. When I started doing anomaly detection and analysis in 2003, I tried hands-on learning of Prolog (for memory dump analysis inference), expert systems (for software support), became familiar with neural networks (C++ implementation at that time). However, my own explorative data analysis approaches, especially for traces and logs, pushed me into human learning, and only recently, I caught up with the latest frameworks and approaches in machine learning. In November 2018, I invented a baseplate representation of chemical structures using LEGO®, and in January 2020, I got an idea to represent clustering using bricks. In the previous years, I used bricks to represent some simple data structures and even software logs, so all that fused into these series of short books (epochs) you are reading now.

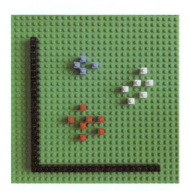

For this epoch, I used the following books as a reference and inspiration:

- Neural Networks: A Systematic Introduction by Raúl Rojas
- Introduction to Deep Learning: From Logical Calculus to Artificial Intelligence by Sandro Skansi
- Artificial Intelligence Engines: A Tutorial Introduction to the Mathematics of Deep Learning by James V. Stone

Before we delve into hands-on internals of linear associative networks, we represent a 2-layer network with 3 neurons each and 9 interconnections, each having a different weight.

We start with the simplest neural network that has two layers with one neuron each. A connection between them is represented with an arrow with its middle part depicting the weight.

Algebraic variables associated with neurons and weights.

x * w = y

$$y = f(x) = wx$$

Input layer

w

Output layer

During actual network operation, input values are assigned to variables representing input neurons. The network processes input using weight values and stores output values in output neurons.

a ∗ = b

$x = a$ w $y = wa$

$b = f(a) = wa$

Input value   Output value

However, the output value may not be the one we expect.

If we expect the neural network to produce output value c on input value a, we need to find the correct value of the connecting weight.

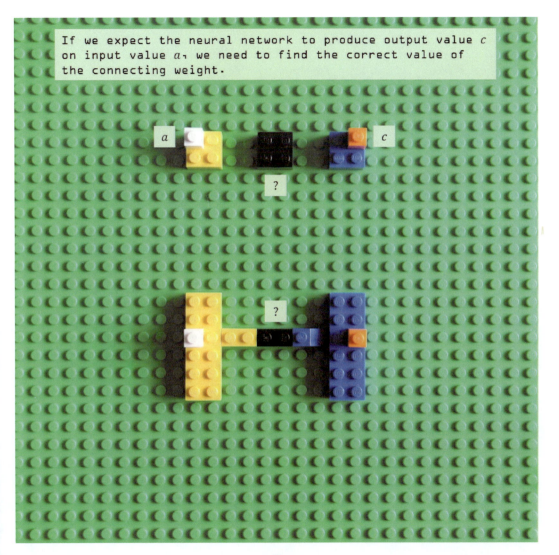

The procedure to find the correct weights (or a function with parameters, called **model**, in general) is called the **training** of the neural network. The goal is to find the weight w' that produces the closest output to the known value c.

a

w'

c

$$c \cong f'(a) = w'a$$

Usually, this procedure is iterative. We get close and close to the correct weight with each iteration (**epoch**).

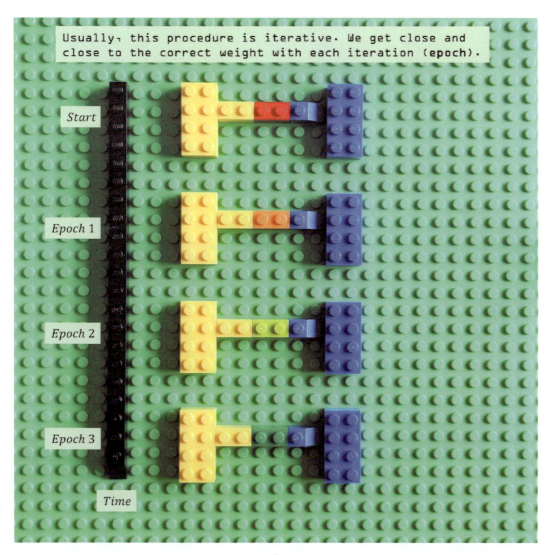

Start

Epoch 1

Epoch 2

Epoch 3

Time

Before we look at calculations needed to find the best weights, we need to depict algebraic operations. Brick expressions are read from bottom to top.

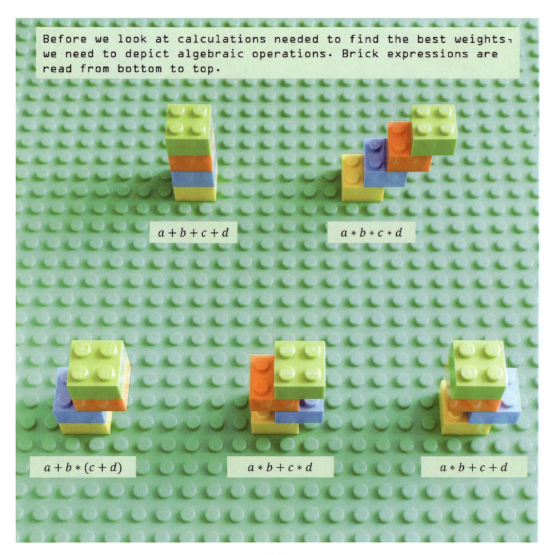

$a+b+c+d$

$a*b*c*d$

$a+b*(c+d)$

$a*b+c*d$

$a*b+c+d$

When we have ambiguities, we can introduce brackets as thin bricks. But we aim to reduce the number of such extra bricks.

$a + b * c$

$a * (b + c)$

$(a + b) * c$

$a * b + c$

For subtraction, we use white bricks placed
underneath the subtracted variable or value.

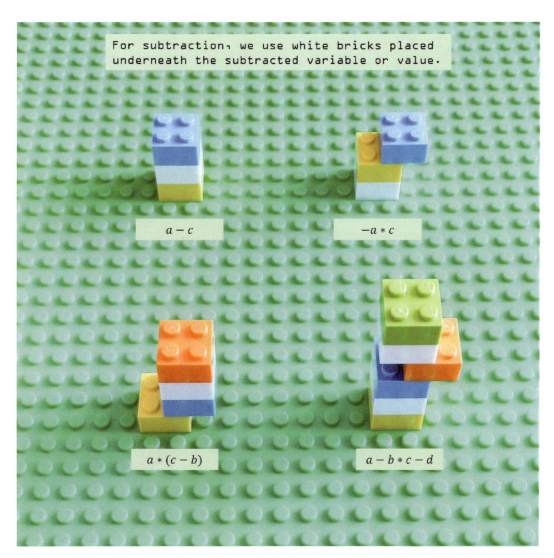

$a - c$

$-a * c$

$a * (c - b)$

$a - b * c - d$

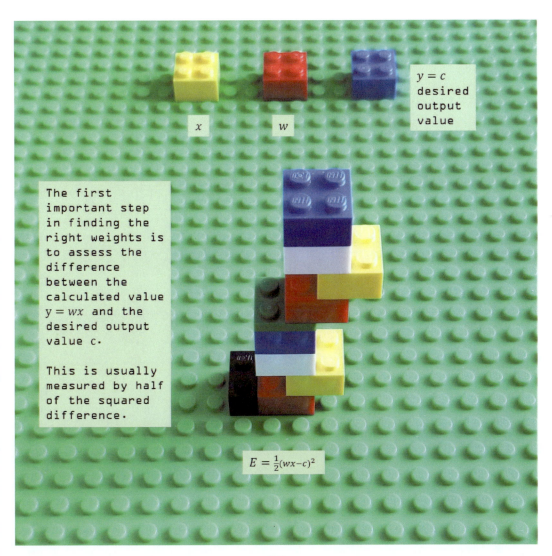

x

w

$y = c$
desired
output
value

The first
important step
in finding the
right weights is
to assess the
difference
between the
calculated value
$y = wx$ and the
desired output
value c.

This is usually
measured by half
of the squared
difference.

$$E = \tfrac{1}{2}(wx-c)^2$$

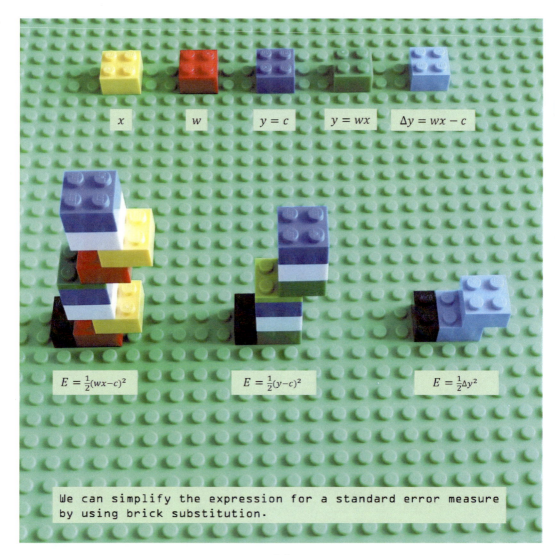

x

w

$y = c$

$y = wx$

$\Delta y = wx - c$

$E = \frac{1}{2}(wx-c)^2$

$E = \frac{1}{2}(y-c)^2$

$E = \frac{1}{2}\Delta y^2$

We can simplify the expression for a standard error measure by using brick substitution.

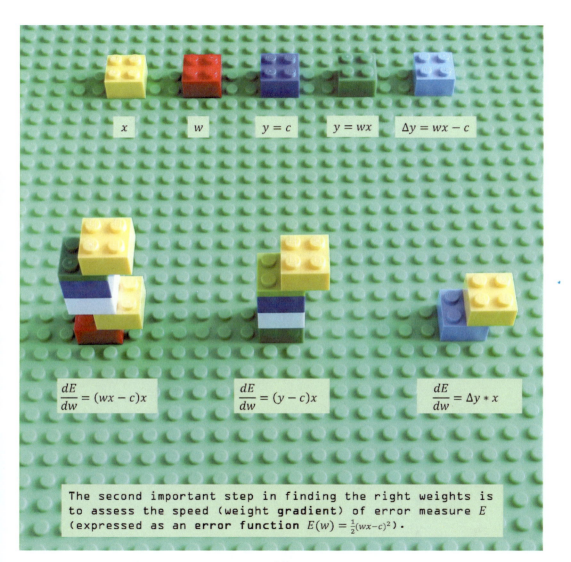

x

w

$y = c$

$y = wx$

$\Delta y = wx - c$

$\dfrac{dE}{dw} = (wx - c)x$

$\dfrac{dE}{dw} = (y - c)x$

$\dfrac{dE}{dw} = \Delta y * x$

The second important step in finding the right weights is to assess the speed (weight gradient) of error measure E (expressed as an error function $E(w) = \frac{1}{2}(wx-c)^2$).

15

x

w

$y = c$

$y = wx$

$\Delta y = wx - c$

w'

w

$$\Delta w = -\varepsilon \frac{dE}{dw} = -\varepsilon \Delta y x$$

The third step of this procedure is adjusting the weight w by a small Δw to get the new weight w'. This adjustment depends on the learning rate ε (**hyperparameter**). We use $-$ sign because we want to reduce the error.

We now update the connection weight.

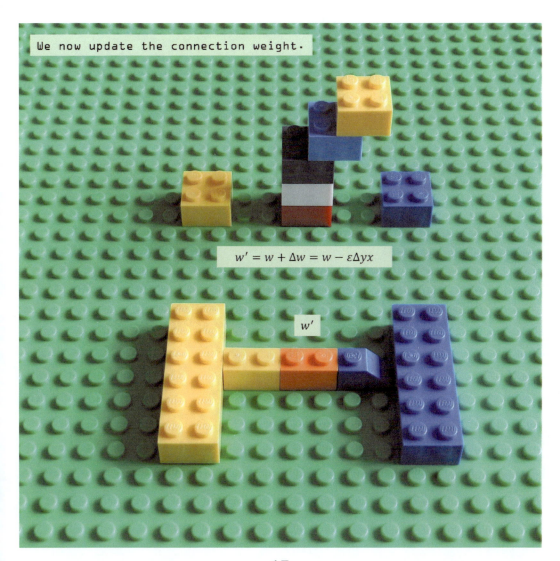

$$w' = w + \Delta w = w - \varepsilon \Delta y x$$

w'

Gradient descent: we repeat the procedure of adjusting weights until the error function is not changing much (gradient is close to 0).

Start

$$\frac{dE}{dw} \not\equiv 0$$

Epoch 1

$$\frac{dE}{dw} \not\equiv 0$$

Epoch 2

$$\frac{dE}{dw} \not\equiv 0$$

Epoch 3

$$\frac{dE}{dw} \cong 0$$

Time

www.ingramcontent.com/pod-product-compliance
Lightning Source LLC
LaVergne TN
LVRC080859070326
832902LV00005B/89

9 781912 636501